# γ Gamma

## Single and Multiple-Digit Multiplication

## Tests

# Math·U·See®

1-888-854-MATH (6284) - mathusee.com

sales@mathusee.com

**Gamma Tests: Single and Multiple-Digit Multiplication**

©2012 Math-U-See, Inc.
Published and distributed by Demme Learning

**mathusee.com**

1-888-854-6284 or +1 717-283-1448 | demmelearning.com
Lancaster, Pennsylvania USA

ISBN 978-1-60826-074-4
Revision Code 1118-B

Printed in the United States of America by CJK Group
 2 3 4 5 6 7 8 9 10

For information regarding CPSIA on this printed material call: 1-888-854-6284
and provide reference #1118-02032020

# LESSON TEST

Fill in the parentheses with the factors and write the product in the oval. Then write the problem two ways beside the rectangle.

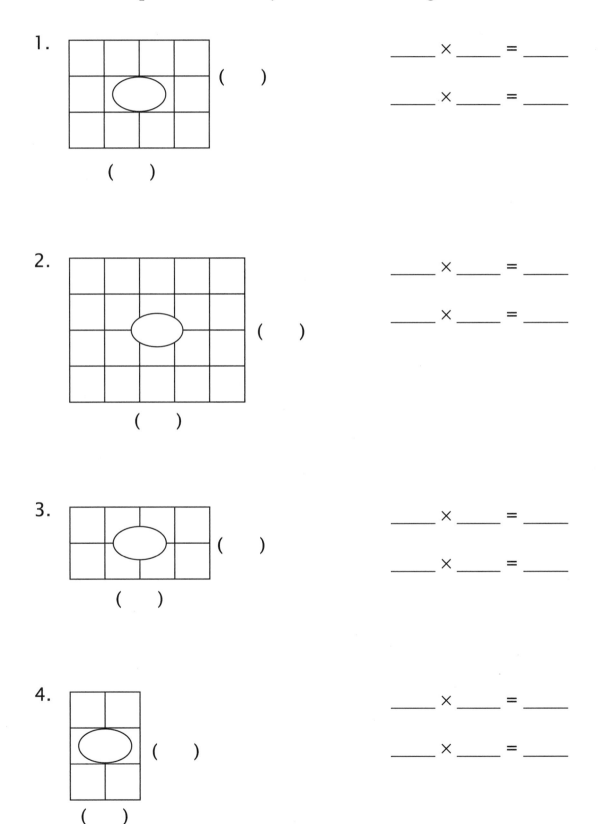

1.

( )

( )

_____ × _____ = _____

_____ × _____ = _____

2.

( )

( )

_____ × _____ = _____

_____ × _____ = _____

3.

( )

( )

_____ × _____ = _____

_____ × _____ = _____

4.

( )

( )

_____ × _____ = _____

_____ × _____ = _____

5.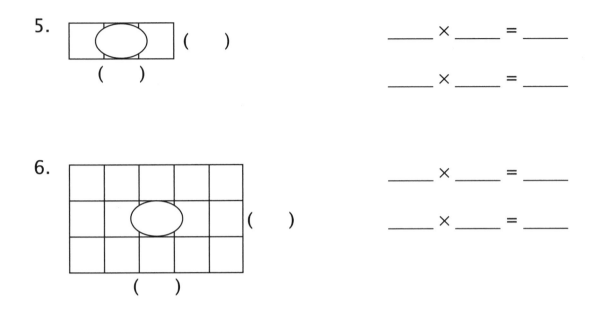

_____ × _____ = _____

_____ × _____ = _____

6.

_____ × _____ = _____

_____ × _____ = _____

7. Draw the rectangle that is the same as:

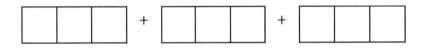

8. Draw the rectangle that is the same as:

## LESSON TEST

Find the answer by multiplying.

1. $0 \times 2 = $ _____

2. $9 \times 1 = $ _____

3. $1 \times 7 = $ _____

4. $5 \times 1 = $ _____

5. $(0)(9) = $ _____

6. $(3)(0) = $ _____

7. $(0)(7) = $ _____

8. $(2)(1) = $ _____

9. $1 \cdot 6 = $ _____

10. $1 \cdot 1 = $ _____

11. $0 \cdot 1 = $ _____

12. $8 \cdot 1 = $ _____

13. $1 \times 3 = $ _____

14. $(6)(0) = $ _____

15. $(0)(5) = $ _____

16. $4 \cdot 0 = $ _____

17.  4
    $\times\,1$

18.  8
    $\times\,0$

19.  1
    $\times\,9$

20.  1
    $\times\,5$

21. How much is five counted one time?

22. How much is zero counted four times?

23. How much is seven counted one time?

24. Carolyn has one candy dish. In the dish are six pieces of candy. How many pieces of candy does Carolyn have in all?

25. Paige counted the noses of all the children on the bus and counted 10. If each child on the bus had one nose, how many children were on the bus?

# LESSON TEST

Skip count and write the numbers.

1. 2, _____, _____, _____, _____, _____, _____, _____, 18, _____

2. 5, _____, _____, _____, _____, _____, _____, 40, _____, _____

3. 10, _____, 30, _____, _____, _____, _____, _____, _____, _____

Find the answer by multiplying.

4. 8 · 1 = _____

5. 0 × 5 = _____

6. 2 · 1 = _____

7. (9)(0) = _____

8.    1
   × 7
   ____

9.    1
   × 0
   ____

10.    6
    × 1
    ____

11.    4
    × 0
    ____

Add or subtract.

12.
$$\begin{array}{r} 5 \\ + 7 \\ \hline \end{array}$$

13.
$$\begin{array}{r} 1\ 4 \\ -\ 5 \\ \hline \end{array}$$

14.
$$\begin{array}{r} 1\ 3 \\ -\ 6 \\ \hline \end{array}$$

15.
$$\begin{array}{r} 7 \\ + 8 \\ \hline \end{array}$$

16. How many is zero counted five times?

17. How many is one counted eight times?

18. Emily read five books a day for three days. Skip count to find how many books she read in all.

19. Chad has four pockets. He put 10 jelly beans in each pocket. How many jelly beans are in his pockets? Skip count to find the answer.

20. Daniel eats two apples a day. How many apples will he eat in eight days? (skip count)

## LESSON TEST

Find the answer by multiplying.

1.  $9 \times 2 =$ _____

2.  $2 \times 4 =$ _____

3.  $(1)(2) =$ _____

4.  $(2)(3) =$ _____

5.  $(6)(2) =$ _____

6.  $(2)(5) =$ _____

7.  $7 \cdot 2 =$ _____

8.  $2 \cdot 2 =$ _____

9.
$$\begin{array}{r} 8 \\ \times\ 2 \\ \hline \end{array}$$

10.
$$\begin{array}{r} 6 \\ \times\ 0 \\ \hline \end{array}$$

11.
$$\begin{array}{r} 1\ 0 \\ \times\ 2 \\ \hline \end{array}$$

12.
$$\begin{array}{r} 3 \\ \times\ 1 \\ \hline \end{array}$$

Skip count and write the numbers.

13.  _____, _____, 15, _____, _____, _____, _____, _____, _____, _____

Add or subtract.

14.
$$\begin{array}{r} 9 \\ -\ 4 \\ \hline \end{array}$$

15.
$$\begin{array}{r} 5 \\ +\ 3 \\ \hline \end{array}$$

16.
$$\begin{array}{r} 1\ 7 \\ -\ 7 \\ \hline \end{array}$$

17.
$$\begin{array}{r} 9 \\ +\ 8 \\ \hline \end{array}$$

18. Rewrite using place-value notation:

263 = _____ + _____ + _____

19. Hazel put six quarts of strawberry jam into pint jars. How many jars did she fill?

20. Naomi made four sandwiches. She used two pieces of bread in each sandwich. How many pieces of bread did Naomi use in all?

## LESSON TEST

Find the answer by multiplying.

1. $2 \times 10 =$ _____

2. $10 \times 9 =$ _____

3. $3 \times 10 =$ _____

4. $10 \times 7 =$ _____

5. $(6)(10) =$ _____

6. $(10)(1) =$ _____

7. $4 \cdot 10 =$ _____

8. $10 \cdot 5 =$ _____

9.
$$\begin{array}{r} 1\,0 \\ \times\ 8 \\ \hline \end{array}$$

10.
$$\begin{array}{r} 5 \\ \times\ 2 \\ \hline \end{array}$$

11.
$$\begin{array}{r} 1 \\ \times\ 3 \\ \hline \end{array}$$

12.
$$\begin{array}{r} 8 \\ \times\ 2 \\ \hline \end{array}$$

Add or subtract.

13.
$$\begin{array}{r} 3\,4 \\ -\ 2\,1 \\ \hline \end{array}$$

14.
$$\begin{array}{r} 5\,5 \\ +\ 4\,2 \\ \hline \end{array}$$

15.  $\begin{array}{r} 1\,8 \\ -\ \ 1 \\ \hline \end{array}$       16.  $\begin{array}{r} 6\,0 \\ +\,1\,7 \\ \hline \end{array}$

17.  Rewrite using place-value notation:

$194 = \underline{\hspace{1.5cm}} + \underline{\hspace{1.5cm}} + \underline{\hspace{1.5cm}}$

18.  Jeremy has seven dimes. How many cents does he have?

19.  Christa bought 10 quarts of milk. How many pints of milk did she buy?

Her son and his friends drank 10 pints of milk. How many pints were left over?

20.  Jason jogged 3 miles a day for 10 days. How many miles did he jog altogether?

## LESSON TEST

Find the answer by multiplying.

1. $3 \times 5 =$ _____

2. $5 \times 7 =$ _____

3. $2 \times 5 =$ _____

4. $5 \times 8 =$ _____

5. $(6)(5) =$ _____

6. $(5)(0) =$ _____

7. $10 \cdot 5 =$ _____

8. $5 \cdot 5 =$ _____

9.
$$\begin{array}{r} 5 \\ \times\ 4 \\ \hline \end{array}$$

10.
$$\begin{array}{r} 2 \\ \times\ 7 \\ \hline \end{array}$$

11.
$$\begin{array}{r} 9 \\ \times\ 2 \\ \hline \end{array}$$

12.
$$\begin{array}{r} 1\,0 \\ \times\ 6 \\ \hline \end{array}$$

Regroup and add.

13.
$$\begin{array}{r} 6\,1 \\ +\ 3\,9 \\ \hline \end{array}$$

14.
$$\begin{array}{r} 4\,7 \\ +\ 2\,5 \\ \hline \end{array}$$

15.  5 6
   + 3 6

16.  8 4
   + 1 9

17. What is five counted eight times?

18. Madelyn did two chores a day for four days. Then she did five chores a day for three days. How many chores did Madelyn do in all?

19. Justin has three nickels and eight dimes. How much money does he have in all?

20. Charlotte sang six songs. If each song was five minutes long, for how long did Charlotte sing?

# UNIT TEST **Lessons 1-6**

## Find the answer by multiplying.

1. $2 \times 2 =$ _____

2. $7 \times 10 =$ _____

3. $10 \times 10 =$ _____

4. $5 \times 2 =$ _____

5. $0 \times 1 =$ _____

6. $10 \times 4 =$ _____

7. $2 \times 8 =$ _____

8. $1 \times 5 =$ _____

9. $(5)(9) =$ _____

10. $(4)(5) =$ _____

11. $1 \cdot 6 =$ _____

12. $3 \cdot 5 =$ _____

13. $(1)(9) =$ _____

14. $(10)(2) =$ _____

15. $3 \cdot 10 =$ _____

16. $5 \cdot 8 =$ _____

17.
$$\begin{array}{r} 0 \\ \times\, 0 \\ \hline \end{array}$$

18.
$$\begin{array}{r} 1\,0 \\ \times\, 6 \\ \hline \end{array}$$

19.   6
    × 5

20.   2
    × 7

21.   3
    × 0

22.   2
    × 3

23.   7
    × 5

24.   2
    × 9

25.   6
    × 2

26.  1 0
    × 8

27.   4
    × 2

28.   5
    × 5

29.  1 0
    × 9

30.   0
    × 2

31.   7
    × 1

32.   1 0
    × 5

Fill in the blanks.

33.  1 quart = _____ pints

34.  1 dime = _____ cents

35.  1 nickel = _____ cents

Find the product, or area, and write it on the line under each drawing.
Label each answer.

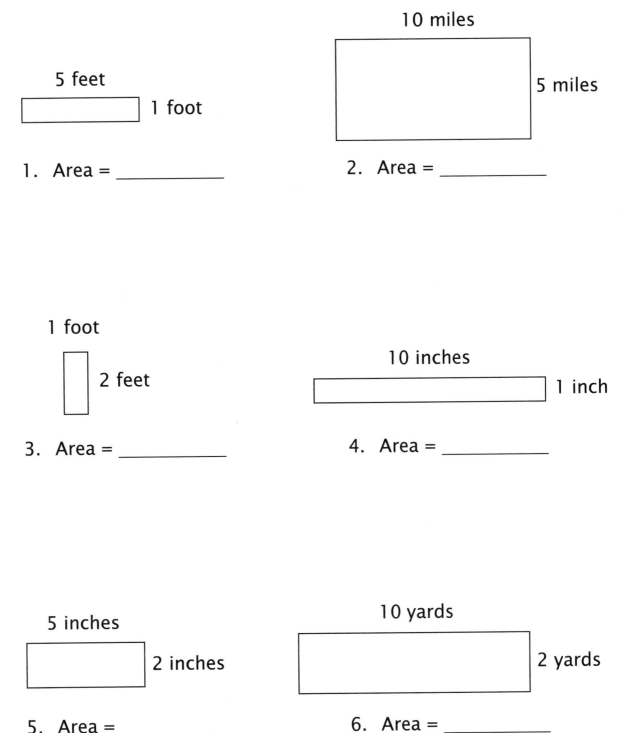

5 feet

1 foot

1. Area = _____

10 miles

5 miles

2. Area = _____

1 foot

2 feet

3. Area = _____

10 inches

1 inch

4. Area = _____

5 inches

2 inches

5. Area = _____

10 yards

2 yards

6. Area = _____

7. $5 \cdot 6 =$ _____

8. $2 \times 8 =$ _____

9. $10 \cdot 7 =$ _____

10. $(5)(5) =$ _____

11. Shauna cut out a piece of wrapping paper that measured two feet by four feet. What was the area of her piece of wrapping paper?

12. Reneé had an aquarium that was three feet long and two feet wide. What was the area of the surface of the water in the aquarium?

# LESSON TEST

Solve for the unknown.

1.  10G = 20

2.  3H = 3

3.  9L = 45

4.  5Z = 50

5.  2Y = 8

6.  5Z = 10

7.  2T = 18

8.  7X = 14

Find the area of each rectangle or square.

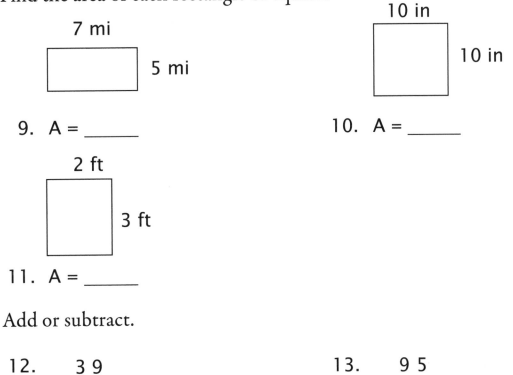

9.  A = _____

10.  A = _____

11.  A = _____

Add or subtract.

12.
```
   3 9
 + 5 2
```

13.
```
   9 5
 - 1 6
```

14.  78
   + 25
   —————

15.  61
   − 48
   —————

16. Rick drove 45 miles on Monday and 37 miles on Tuesday. How many more miles did he drive on Monday than on Tuesday?

17. Joe needs 40 gallons of water. If he carries five gallons per trip, how many trips will he need to take?

18. When Anne changed the oil in her car, she used 12 pints of oil. How many quarts is that?

19. Kayla bought $35 worth of books. How many five-dollar bills did she need to pay for the books?

20. If Bigfoot goes 10 feet per step, how many steps must he take to cover 50 feet?

# 9

Skip count and write the numbers.

1. _____, _____, 27, _____, _____, _____, _____, _____, _____, _____

Find the missing multiples of 2 and 9 in the equivalent fractions.

2. $\dfrac{2}{9} = \dfrac{\phantom{6}}{\phantom{6}} = \dfrac{6}{\phantom{6}} = \dfrac{\phantom{6}}{36} = \dfrac{\phantom{6}}{\phantom{6}} = \dfrac{\phantom{6}}{54} = \dfrac{\phantom{6}}{\phantom{6}} = \dfrac{16}{\phantom{6}} = \dfrac{\phantom{6}}{\phantom{6}} = \dfrac{\phantom{6}}{\phantom{6}}$

Skip count and write the missing numbers. Then fill in the missing factors under the lines.

3. $\dfrac{0}{(5)(0)}$   $\dfrac{5}{(5)(\ \ )}$   $\dfrac{\phantom{5}}{(5)(\ \ )}$   $\dfrac{\phantom{5}}{(5)(3)}$   $\dfrac{20}{(5)(\ \ )}$   $\dfrac{\phantom{5}}{(5)(5)}$

$\dfrac{\phantom{5}}{(5)(\ \ )}$   $\dfrac{\phantom{5}}{(5)(7)}$   $\dfrac{40}{(5)(\ \ )}$   $\dfrac{\phantom{5}}{(5)(9)}$   $\dfrac{\phantom{5}}{(5)(10)}$

Solve for the unknown.

4. $5J = 45$

5. $8C = 16$

6. $5N = 35$

7. $8R = 80$

Find the area of each rectangle or square.

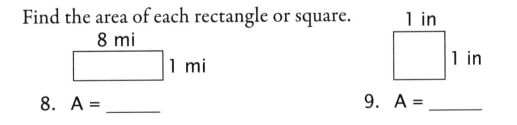

8. A = _____

9. A = _____

10. A = _____

Add or subtract.

11.  $\begin{array}{r} 3\ 2 \\ +\ 5\ 8 \\ \hline \end{array}$

12.  $\begin{array}{r} 5\ 4 \\ -\ 4\ 7 \\ \hline \end{array}$

13.  $\begin{array}{r} 6\ 5 \\ +\ 1\ 8 \\ \hline \end{array}$

14.  $\begin{array}{r} 7\ 0 \\ -\ 2\ 1 \\ \hline \end{array}$

15. Caleb has four dimes and three nickels. How much money does he have altogether?

16. Marvin has three bags. Each bag holds nine marbles. How many marbles does he have altogether?

17. Shelley spent $9 apiece for five books. How much did she spend altogether?

18. Each cookie contains nine chocolate chips. How many chips are in nine cookies?

Find the answer by multiplying.

1.  $9 \times 0 =$ _____

2.  $9 \times 9 =$ _____

3.  $3 \times 9 =$ _____

4.  $7 \times 9 =$ _____

5.  $(9)(4) =$ _____

6.  $(6)(9) =$ _____

7.  $9 \cdot 8 =$ _____

8.  $2 \cdot 9 =$ _____

9.  $\begin{array}{r} 1\,0 \\ \times\ 9 \\ \hline \end{array}$

10.  $\begin{array}{r} 9 \\ \times\ 5 \\ \hline \end{array}$

11.  $\begin{array}{r} 2 \\ \times\ 6 \\ \hline \end{array}$

12.  $\begin{array}{r} 5 \\ \times\ 8 \\ \hline \end{array}$

Skip count and write the missing numbers. Then fill in the missing factors under the lines.

13.
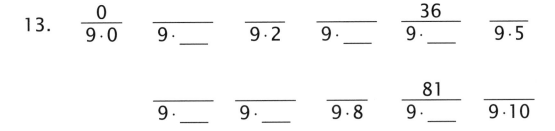

Solve for the unknown.

14. $2X = 18$

15. $5T = 25$

16. $4H = 0$

17. $8Y = 72$

18. Andrea made three phone calls. Each call lasted nine minutes. How many minutes did Andrea spend on the phone?

19. If Jenny's nine cats really have nine lives each, how many lives do they have in all?

20. Roy spent $9 per week on gasoline. How much did he spend in eight weeks?

# LESSON TEST

Skip count and write the numbers.

1. _____, 6, _____, _____, _____, _____, _____, _____, _____, _____

Find the missing multiples of 3 and 5 in the equivalent fractions.

2. $\dfrac{3}{5} = \dfrac{}{} = \dfrac{}{20} = \dfrac{}{} = \dfrac{}{30} = \dfrac{}{} = \dfrac{24}{} = \dfrac{}{} = \dfrac{}{}$

Skip count and write the missing numbers. Then fill in the missing factors under the lines.

3.
$$\frac{0}{9 \times 0} \qquad \frac{\phantom{00}}{9 \times \underline{\phantom{0}}} \qquad \frac{18}{9 \times \underline{\phantom{0}}} \qquad \frac{\phantom{00}}{9 \times 3} \qquad \frac{36}{9 \times \underline{\phantom{0}}} \qquad \frac{\phantom{00}}{9 \times \underline{\phantom{0}}}$$

$$\frac{\phantom{00}}{9 \times 6} \qquad \frac{\phantom{00}}{9 \times \underline{\phantom{0}}} \qquad \frac{72}{9 \times \underline{\phantom{0}}} \qquad \frac{\phantom{00}}{9 \times 9} \qquad \frac{\phantom{00}}{9 \times \underline{\phantom{0}}}$$

Multiply.

4. $(5)(4) =$ _____

5. $2 \times 7 =$ _____

6. $5 \times 3 =$ _____

7. $2 \cdot 8 =$ _____

Add. Make ten when possible.

8.
$$\begin{array}{r} 2\;1 \\ 3\;9 \\ +\;\;\;6 \\ \hline \end{array}$$

9.
$$\begin{array}{r} 2\;8 \\ 4\;0 \\ +\;6\;1 \\ \hline \end{array}$$

10.
```
   6 5
   2 3
   1 5
+   7
```

11.
```
   3 2
   3 3
   4 2
+ 1 3
```

Use skip counting to solve the word problems.

12. Each of Shane's four horses won three ribbons. How many ribbons did they win in all?

13. Kerri ate three meals a day for a week. How many meals did she eat that week?

14. Vernon wrote letters to five of his friends. Each letter was three pages long. How many pages did he write altogether?

15. Edith ran three miles a day for six days. How far did she run during that time?

## LESSON TEST

Find the answer by multiplying.

1. $3 \times 9 = $ _____

2. $8 \times 3 = $ _____

3. $3 \cdot 3 = $ _____

4. $4 \cdot 3 = $ _____

5. $3 \times 2 = $ _____

6. $6 \times 3 = $ _____

7. $(3)(10) = $ _____

8. $(7)(3) = $ _____

9. $\begin{array}{r} 3 \\ \times\,1 \\ \hline \end{array}$

10. $\begin{array}{r} 5 \\ \times\,3 \\ \hline \end{array}$

11. $\begin{array}{r} 6 \\ \times\,9 \\ \hline \end{array}$

12. $\begin{array}{r} 9 \\ \times\,9 \\ \hline \end{array}$

Add or subtract.

13. $\begin{array}{r} 4\,2 \\ 3\,4 \\ +\quad 8 \\ \hline \end{array}$

14. $\begin{array}{r} 1\,7 \\ 1\,1 \\ +1\,3 \\ \hline \end{array}$

Add or subtract.

15.     9 2
      – 2 5

16.     8 4
      – 3 6

Skip count or multiply by three to find the number of feet. Use the symbol for feet in your answer.

17.
| 1 ft | 1 ft | 1 ft | 1 ft | 1 ft | 1 ft | 1 ft | 1 ft | 1 ft |
| 1 ft | 1 ft | 1 ft | 1 ft | 1 ft | 1 ft | 1 ft | 1 ft | 1 ft |

6 yards = _____

Skip count or multiply by three to find the number of teaspoons. Each picture represents a tablespoon.

18.

9 tablespoons = _____ teaspoons

19. Eric carried the football for nine yards before being tackled. For how many feet did he carry the ball?

20. Mrs. Miller made three sandwiches for each of her four children. How many sandwiches did Mrs. Miller make in all?

Skip count and write the numbers.

1. ____, 12, ____, ____, ____, ____, ____, ____, ____, ____

2. ____, 18, ____, ____, ____, ____, ____, ____, ____, ____

Fill in the blanks with the correct numerators and denominators to name the equivalent fractions.

3.

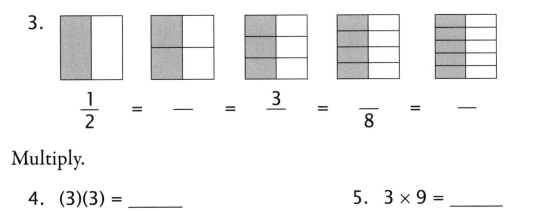

$$\frac{1}{2} = \frac{\phantom{0}}{\phantom{0}} = \frac{3}{\phantom{0}} = \frac{\phantom{0}}{8} = \frac{\phantom{0}}{\phantom{0}}$$

Multiply.

4. (3)(3) = _____

5. 3 × 9 = _____

6. 9 × 9 = _____

7. 0 · 0 = _____

Find the perimeter of each rectangle or square.

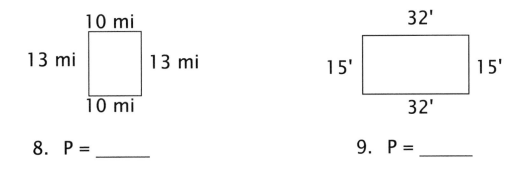

8. P = _____

9. P = _____

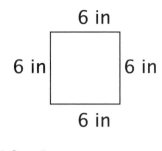

10. P = _____

Use skip counting to solve the word problems.

11. Benjamin bought five packs of light bulbs. There are six bulbs in a pack. How many light bulbs did he buy in all?

12. Miguel wrote six letters every day for a week. How many letters did he write in all?

13. Myles arranged some blocks into rows. If he made four rows with six in a row, how many blocks did he use?

14. Liane's family made six trips every year to visit her grandparents. How many trips did they make over a period of six years?

## LESSON TEST

Find the answer by multiplying.

1. $6 \times 7 =$ _____

2. $9 \times 6 =$ _____

3. $6 \times 3 =$ _____

4. $1 \times 6 =$ _____

5. $(6)(4) =$ _____

6. $(10)(6) =$ _____

7. $(6)(8) =$ _____

8. $(6)(6) =$ _____

9. $6 \cdot 2 =$ _____

10. $3 \cdot 3 =$ _____

11. $6 \cdot 5 =$ _____

12. $2 \cdot 10 =$ _____

13.
$$\begin{array}{r} 5 \\ \times\ 5 \\ \hline \end{array}$$

14.
$$\begin{array}{r} 7 \\ \times\ 9 \\ \hline \end{array}$$

15.
$$\begin{array}{r} 9 \\ \times\ 9 \\ \hline \end{array}$$

16.
$$\begin{array}{r} 8 \\ \times\ 3 \\ \hline \end{array}$$

Fill in the blanks with the correct numerators and denominators to name the equivalent fractions.

17.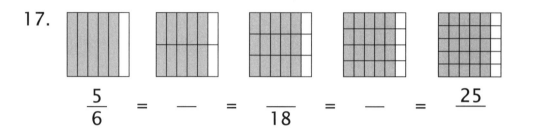

$$\frac{5}{6} = \frac{}{} = \frac{}{18} = \frac{}{} = \frac{25}{}$$

18. Mom said that Clara could eat two tablespoons of peanut butter. How many teaspoons of peanut butter may she eat?

19. Ben needs six yards of rope. How many feet of rope should he buy?

20. Diane uses six cups of flour in every batch of bread that she makes. How many cups of flour does she need for six batches of bread?

Find the answer by multiplying.

1.  $8 \times 6 = $ _____

2.  $3 \times 5 = $ _____

3.  $9 \times 6 = $ _____

4.  $3 \times 7 = $ _____

5.  $1 \times 9 = $ _____

6.  $6 \times 0 = $ _____

7.  $10 \times 9 = $ _____

8.  $9 \times 8 = $ _____

9.  $(3)(2) = $ _____

10.  $(7)(9) = $ _____

11.  $3 \cdot 1 = $ _____

12.  $3 \cdot 6 = $ _____

13.  $(10)(3) = $ _____

14.  $(6)(3) = $ _____

15.  $4 \cdot 9 = $ _____

16.  $6 \cdot 4 = $ _____

17.  $\begin{array}{r} 2 \\ \times\ 6 \\ \hline \end{array}$

18.  $\begin{array}{r} 3 \\ \times\ 3 \\ \hline \end{array}$

19.    5
    × 6

20.    9
    × 2

21.    3
    × 9

22.    9
    × 9

23.    5
    × 9

24.    3
    × 8

25.    4
    × 3

26.    6
    × 6

27.    7
    × 6

28.   1 0
    × 6

Answer the questions.

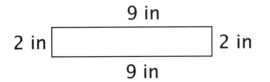

29. What is the area of the rectangle?

30. What is the perimeter of the rectangle?

31. Carlos has a sailboat that is four yards long. How long is his sailboat in feet?

32. Julia's recipe calls for six tablespoons of lemon juice. How many teaspoons does she need?

# LESSON TEST

Skip count and write the numbers.

1. _____, _____, _____, 16, _____, _____, _____, _____, _____, _____

2. _____, _____, _____, 12, _____, _____, _____, _____, _____, _____

Fill in the blanks with the correct numerators and denominators to name the equivalent fractions.

3.

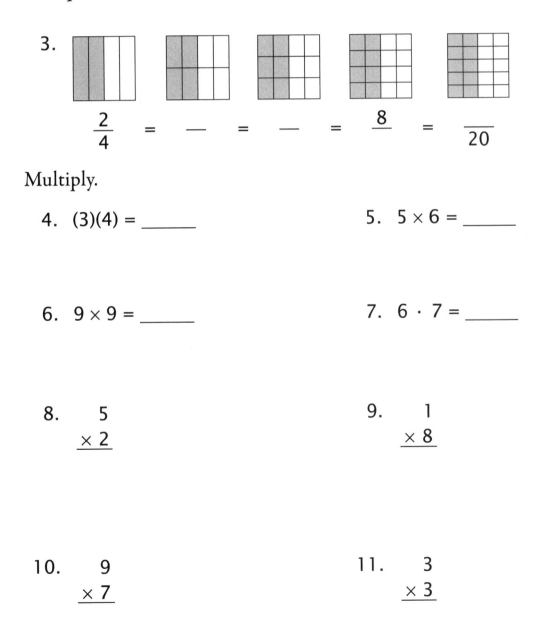

$$\frac{2}{4} = \frac{}{\phantom{-}} = \frac{}{\phantom{-}} = \frac{8}{\phantom{-}} = \frac{}{20}$$

Multiply.

4. (3)(4) = _____

5. 5 × 6 = _____

6. 9 × 9 = _____

7. 6 · 7 = _____

8.  5
   × 2

9.  1
   × 8

10.  9
    × 7

11.  3
    × 3

Skip count by four to find how many quarts there are in all.

12.

Use skip counting to solve the word problems.

13. It takes Lindsay four minutes to ride a mile on her bike. How long does it take her to ride three miles?

14. Joanne slides down the snowy hill four times every hour. How many times does she slide down in four hours?

15. Lyle watched seven horses gallop across the field. How many hooves were making tracks?

## LESSON TEST

Find the answer by multiplying.

1. $4 \times 9 =$ _____

2. $4 \times 4 =$ _____

3. $4 \cdot 2 =$ _____

4. $5 \cdot 4 =$ _____

5. $4 \times 3 =$ _____

6. $8 \times 4 =$ _____

7. $(4)(7) =$ _____

8. $(4)(6) =$ _____

9.  $\begin{array}{r} 10 \\ \times\ 4 \\ \hline \end{array}$

10.  $\begin{array}{r} 9 \\ \times\ 6 \\ \hline \end{array}$

11.  $\begin{array}{r} 3 \\ \times\ 7 \\ \hline \end{array}$

12.  $\begin{array}{r} 6 \\ \times\ 6 \\ \hline \end{array}$

Find the perimeter and area of the square.

13. Area = _____

14. Perimeter = _____

5 in

5 in ☐ 5 in

5 in

Solve for the unknown and fill in the blanks.

15. 12 pints = _____ quarts

16. 9 feet = _____ yards

17. 30 cents = _____ dimes

Multiply by four to find how many quarters have the same value as six dollars.

18.

$4 \times 6 =$ _____

19. Joseph's bucket holds five gallons of water. He uses one quart of water for each of his tomato plants. How many tomato plants can he water with one bucketful?

20. Gumballs cost a quarter apiece. How many gumballs can Sam buy for three dollars?

# LESSON TEST

Skip count and write the missing numbers.

1. _____, 14, _____, _____, _____, _____, _____, _____, _____, _____

Fill in the blanks with the correct numerators and denominators to name the equivalent fractions.

2.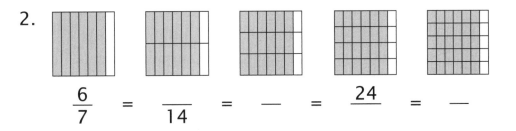

$$\frac{6}{7} = \frac{\phantom{00}}{14} = \frac{\phantom{00}}{\phantom{00}} = \frac{24}{\phantom{00}} = \frac{\phantom{00}}{\phantom{00}}$$

Use "mittens" to help you multiply by multiples of 10.

3.  $\begin{array}{r} 70 \\ \times\ 3 \\ \hline \end{array}$

4.  $\begin{array}{r} 40 \\ \times\ 2 \\ \hline \end{array}$

5.  $\begin{array}{r} 90 \\ \times\ 5 \\ \hline \end{array}$

Solve for the unknown.

6. $7X = 14$

7. $4Y = 36$

8. $6R = 42$

9. $5Q = 15$

Write <, >, or = in the oval.

10. $3 \times 9 \bigcirc 4 \times 7$       11. $8 + 5 \bigcirc 9 + 4$

12. $4 \text{ ft} \bigcirc 10 \text{ yd}$

Use skip counting to solve the word problems.

13. Shirley waited for six weeks for her mail-order package to arrive. For how many days did she wait?

14. Matthew received seven points for each correct answer. How many points did he receive for seven correct answers?

15. Sarah set the table for seven people. She used three pieces of silverware for each person. How many pieces of silverware did she use in all?

## LESSON TEST

Find the answer by multiplying.

1. $7 \times 2 = $ _____

2. $4 \times 7 = $ _____

3. $3 \times 4 = $ _____

4. $6 \cdot 7 = $ _____

5.
$$\begin{array}{r} 7 \\ \times\ 9 \\ \hline \end{array}$$

6.
$$\begin{array}{r} 10 \\ \times\ 7 \\ \hline \end{array}$$

7.
$$\begin{array}{r} 7 \\ \times\ 7 \\ \hline \end{array}$$

8.
$$\begin{array}{r} 5 \\ \times\ 7 \\ \hline \end{array}$$

9.
$$\begin{array}{r} 8 \\ \times\ 7 \\ \hline \end{array}$$

10.
$$\begin{array}{r} 6 \\ \times\ 6 \\ \hline \end{array}$$

11.
$$\begin{array}{r} 50 \\ \times\ 5 \\ \hline \end{array}$$

12.
$$\begin{array}{r} 90 \\ \times\ 2 \\ \hline \end{array}$$

Use "mittens" to help you multiply by multiples of 100.

13.
$$\begin{array}{r} 100 \\ \times\ \ 4 \\ \hline \end{array}$$

14.
$$\begin{array}{r} 400 \\ \times\ \ 2 \\ \hline \end{array}$$

15.
$$\begin{array}{r} 200 \\ \times\ \ 2 \\ \hline \end{array}$$

16
$$\begin{array}{r} 100 \\ \times\ \ 6 \\ \hline \end{array}$$

17. How many quarters have the same value as seven dollars?

18. Rod earned eight dollars for each lawn he mowed. How much money did he earn for mowing seven lawns?

19. Rachel put up a fence at the back of her yard. Each section of her fence was seven feet long. There were six sections. How long was Rachel's fence?

20. Mindy bought seven boxes of paper clips. Each one contained 100 clips. How many paper clips did Mindy have altogether?

## LESSON TEST

Skip count and write the numbers.

1.  ____, 16, ____, ____, ____, ____, ____, ____, ____, ____

2.  ____, 14, ____, ____, ____, ____, ____, ____, ____, ____

Fill in the blanks with the correct numerators and denominators to name the equivalent fractions.

3.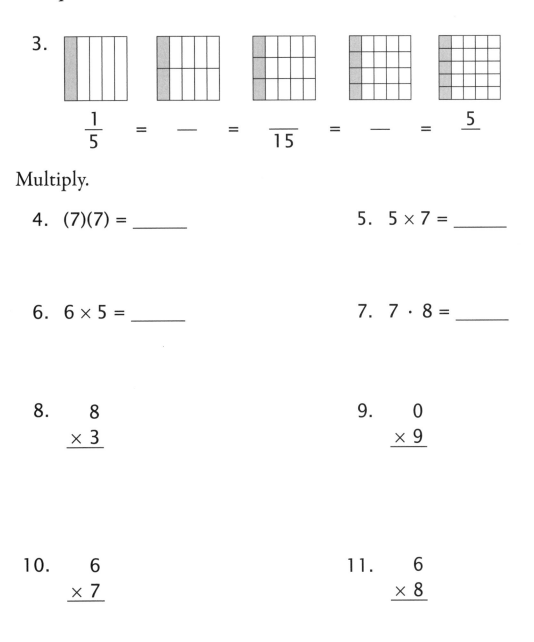

$$\frac{1}{5} = \frac{}{} = \frac{}{15} = \frac{}{} = \frac{5}{}$$

Multiply.

4.  $(7)(7) =$ _____

5.  $5 \times 7 =$ _____

6.  $6 \times 5 =$ _____

7.  $7 \cdot 8 =$ _____

8.  $\begin{array}{r} 8 \\ \times\ 3 \\ \hline \end{array}$

9.  $\begin{array}{r} 0 \\ \times\ 9 \\ \hline \end{array}$

10. $\begin{array}{r} 6 \\ \times\ 7 \\ \hline \end{array}$

11. $\begin{array}{r} 6 \\ \times\ 8 \\ \hline \end{array}$

Add.

12.
```
   4 5 0
 + 1 0 6
```

13.
```
   3 9 2
 + 1 1 7
```

14.
```
   5 4 6
 + 2 3 7
```

Use skip counting to solve the word problems.

15. Miranda bought six gallons of ice cream. Since there are eight pints to a gallon, how many pints did she buy?

16. Frank worked eight hours a day for five days. How many hours did he work altogether?

17. The quilt that Michelle made measured six feet by eight feet. What was the area of the quilt?

18. Elizabeth bought a hat for eight dollars. If she paid for it entirely with quarters, how many quarters did she use?

# LESSON TEST

Find the answer by multiplying.

1.  $8 \times 3 =$ _____

2.  $9 \times 8 =$ _____

3.  $8 \cdot 7 =$ _____

4.  $4 \cdot 8 =$ _____

5.  $8 \times 8 =$ _____

6.  $6 \times 8 =$ _____

7.  $(8)(10) =$ _____

8.  $(5)(8) =$ _____

9.
$$\begin{array}{r} 8\,0 \\ \times\ 2 \\ \hline \end{array}$$

10.
$$\begin{array}{r} 7 \\ \times\ 6 \\ \hline \end{array}$$

11.
$$\begin{array}{r} 5\,0 \\ \times\ 7 \\ \hline \end{array}$$

12.
$$\begin{array}{r} 3\,0\,0 \\ \times\ \ \ 3 \\ \hline \end{array}$$

Write <, >, or = in the oval.

13.  $8 + 1 \bigcirc 8 \times 1$

14.  2 pt $\bigcirc$ 2 gal

15.  $3 \times 8 \bigcirc 6 \times 4$

Subtract.

16.　　　7 2 0
　　　－ 4 3 1

17.　　　8 9 2
　　　－ 6 9 5

18.　　　4 0 6
　　　－ 1 1 9

19. Heidi poured three gallons of fruit punch into glasses that held a pint each. How many glasses did Heidi fill?

20. June learned eight new vocabulary words every day for five days. How many new words did she learn?

Multiply.

1. $4 \times 7 =$ _____

2. $5 \times 5 =$ _____

3. $0 \times 1 =$ _____

4. $8 \times 9 =$ _____

5. $3 \times 4 =$ _____

6. $3 \times 8 =$ _____

7. $5 \times 3 =$ _____

8. $7 \times 10 =$ _____

9. $(3)(3) =$ _____

10. $(10)(6) =$ _____

11. $8 \cdot 5 =$ _____

12. $6 \cdot 4 =$ _____

13. $(8)(2) =$ _____

14. $(4)(9) =$ _____

15. $7 \cdot 7 =$ _____

16. $4 \cdot 5 =$ _____

17.　　6
　　× 1

18.　　7
　　× 5

19.　　3
　　× 6

20.　1 0
　　× 2

21.　　9
　　× 3

22.　1 0
　　× 5

23.　　7
　　× 2

24.　　7
　　× 8

25.　1 0
　　× 3

26.　　4
　　× 2

27.　　8
　　× 8

28.　1 0
　　× 9

29.  $\begin{array}{r} 5 \\ \times\,2 \\ \hline \end{array}$

30.  $\begin{array}{r} 4 \\ \times\,4 \\ \hline \end{array}$

31.  $\begin{array}{r} 8 \\ \times\,6 \\ \hline \end{array}$

32.  $\begin{array}{r} 5 \\ \times\,0 \\ \hline \end{array}$

33.  $\begin{array}{r} 1\,0 \\ \times\,1\,0 \\ \hline \end{array}$

34.  $\begin{array}{r} 6 \\ \times\,9 \\ \hline \end{array}$

35.  $\begin{array}{r} 1\,0 \\ \times\,\ 8 \\ \hline \end{array}$

36.  $\begin{array}{r} 9 \\ \times\,9 \\ \hline \end{array}$

37.  $\begin{array}{r} 6 \\ \times\,5 \\ \hline \end{array}$

38.  $\begin{array}{r} 8 \\ \times\,4 \\ \hline \end{array}$

39.  $\begin{array}{r} 9 \\ \times\,5 \\ \hline \end{array}$

40.  $\begin{array}{r} 6 \\ \times\,6 \\ \hline \end{array}$

41. $2 \times 9 =$ _____

42. $10 \times 4 =$ _____

43. $6 \times 7 =$ _____

44. $2 \times 3 =$ _____

45. $6 \times 2 =$ _____

46. $9 \times 7 =$ _____

47. $2 \times 2 =$ _____

48. $7 \times 3 =$ _____

Fill in the blanks.

49. 5 quarts = _____ pints

50. 8 yards = _____ feet

51. 6 gallons = _____ quarts

52. $10 = _____ quarters

53. 4 Tbsp = _____ tsp

54. 7 gallons = _____ pints

Answer the questions.

55. What is the area of the rectangle?

56. What is the perimeter of the rectangle?

6 ft

5 ft ☐ 5 ft

6 ft

# 21

Multiply using standard notation and place-value notation.

1.  $\begin{array}{r} 2\ 2 \\ \times\ 4 \\ \hline \end{array}$  $\begin{array}{r} 20 + 2 \\ \times\quad 4 \\ \hline \end{array}$     2.  $\begin{array}{r} 1\ 2 \\ \times\ 3 \\ \hline \end{array}$  $\begin{array}{r} 10 + 2 \\ \times\quad 3 \\ \hline \end{array}$

3.  $\begin{array}{r} 1\ 1 \\ \times\ 5 \\ \hline \end{array}$  $\begin{array}{r} 10 + 1 \\ \times\quad 5 \\ \hline \end{array}$     4.  $\begin{array}{r} 4\ 1 \\ \times\ 2 \\ \hline \end{array}$  $\begin{array}{r} 40 + 1 \\ \times\quad 2 \\ \hline \end{array}$

5.  $\begin{array}{r} 2\ 1\ 1 \\ \times\ \ 3 \\ \hline \end{array}$  $\begin{array}{r} 200 + 10 + 1 \\ \times\qquad\qquad 3 \\ \hline \end{array}$     6.  $\begin{array}{r} 2\ 0\ 2 \\ \times\ \ 4 \\ \hline \end{array}$  $\begin{array}{r} 200 + 00 + 2 \\ \times\qquad\qquad 4 \\ \hline \end{array}$

7.  $\begin{array}{r} 1\ 1\ 1 \\ \times\ \ 7 \\ \hline \end{array}$  $\begin{array}{r} 100 + 10 + 1 \\ \times\qquad\qquad 7 \\ \hline \end{array}$     8.  $\begin{array}{r} 1\ 1\ 2 \\ \times\ \ 4 \\ \hline \end{array}$  $\begin{array}{r} 100 + 10 + 2 \\ \times\qquad\qquad 4 \\ \hline \end{array}$

Add or subtract.

9.
```
   2 1
   3 5
 + 2 4
```

10.
```
   4 2
   1 9
 + 3 7
```

11.
```
  2 4 5
 -  1 6
```

12.
```
  3 0 4
 - 2 2 8
```

13. One hundred three people listened to the speech. How many ears were listening?

14. A carton of eggs holds 12 eggs. How many eggs are there in four cartons?

15. Erica rode her bike at 13 miles per hour for 3 hours. How far did she ride in all?

Round to the nearest 10.

1. 26 _____

2. 42 _____

3. 75 _____

Round to the nearest 100.

4. 256 _____

5. 512 _____

6. 128 _____

Round to the nearest 1,000.

7. 6,657 _____

8. 2,585 _____

9. 7,935 _____

Round the top number to the nearest 10 and estimate the answer.

10.　3 7 →
　　 × 3

11.　　7 1 →
　　 ×　2

12.　2 6 →
　　 × 4

Round the top number to the nearest 100 and estimate the answer.

13. 1 9 8 →
    × 7

14. 7 7 3 →
    × 4

15. 2 0 8 →
    × 4

Multiply using standard notation and place-value notation.

16. 2 4    2 0 + 4
    × 2     × 2

17. 1 2    1 0 + 2
    × 3     × 3

18. Mom spent $2,395 on living room furniture. Round that amount to the nearest thousand.

19. Valerie spent $48 a night on a hotel room for six nights. Round the nightly cost to the nearest ten and estimate the total cost of her stay.

20. Aiden drove 213 miles a week on his way to and from work. Round this distance to the nearest hundred and estimate the total distance that he drove in four weeks.

## LESSON TEST

Multiply.

1.
$$\begin{array}{r} 23 \\ \times\,23 \\ \hline \end{array}$$

2.
$$\begin{array}{r} 22 \\ \times\,44 \\ \hline \end{array}$$

3.
$$\begin{array}{r} 21 \\ \times\,11 \\ \hline \end{array}$$

4.
$$\begin{array}{r} 32 \\ \times\,12 \\ \hline \end{array}$$

5.
$$\begin{array}{r} 17 \\ \times\,11 \\ \hline \end{array}$$

6.
$$\begin{array}{r} 18 \\ \times\,11 \\ \hline \end{array}$$

7.
$$\begin{array}{r} 23 \\ \times\,32 \\ \hline \end{array}$$

8.
$$\begin{array}{r} 20 \\ \times\,44 \\ \hline \end{array}$$

Round the top number to the nearest 100 and estimate the answer.

9.
$$\begin{array}{r} 803 \rightarrow \\ \times\quad 5 \\ \hline \end{array}$$

10.
$$\begin{array}{r} 498 \rightarrow \\ \times\quad 3 \\ \hline \end{array}$$

11.
$$\begin{array}{r} 651 \rightarrow \\ \times\quad 4 \\ \hline \end{array}$$

Add.

12.
$$
\begin{array}{r}
1\ 3\ 8 \\
+\ 2\ 7\ 4 \\
\hline
\end{array}
$$

13.
$$
\begin{array}{r}
7\ 0\ 5 \\
+\ 3\ 9\ 8 \\
\hline
\end{array}
$$

14.
$$
\begin{array}{r}
4\ 6\ 4 \\
+\ 1\ 4\ 0 \\
\hline
\end{array}
$$

15. Round 35 to the nearest 10.

16. Round 4,199 to the nearest 1,000.

17. Hannah's bedroom measures 11 feet by 14 feet. What is the area of the room?

18. Kevin bought a dozen muffins a day for two weeks. How many muffins did he buy in all?

Multiply, regrouping if necessary.

1.  $\begin{array}{r} 26 \\ \times 15 \\ \hline \end{array}$

2.  $\begin{array}{r} 31 \\ \times 29 \\ \hline \end{array}$

3.  $\begin{array}{r} 22 \\ \times 35 \\ \hline \end{array}$

4.  $\begin{array}{r} 38 \\ \times 34 \\ \hline \end{array}$

5.  $\begin{array}{r} 28 \\ \times 39 \\ \hline \end{array}$

6.  $\begin{array}{r} 14 \\ \times 16 \\ \hline \end{array}$

7.  $\begin{array}{r} 27 \\ \times 23 \\ \hline \end{array}$

8.  $\begin{array}{r} 16 \\ \times 24 \\ \hline \end{array}$

Round the top number to the nearest 100 and estimate the answer.

9.  $\begin{array}{r} 109 \rightarrow \\ \times \quad 3 \\ \hline \end{array}$

10. $\begin{array}{r} 671 \rightarrow \\ \times \quad 8 \\ \hline \end{array}$

11. $\begin{array}{r} 245 \rightarrow \\ \times \quad 4 \\ \hline \end{array}$

Skip count and write the missing numbers.

12.  _____, _____, _____, _____, _____, _____, _____, _____, _____, 40

Solve for the unknown and fill in the blanks.

13.  50 cents = _____ dimes          14.  32 quarters = $_____

15.  12 ft = _____ yd

16.  Jeffrey drove 48 miles a day for five days. How many miles did he drive in all?

17.  Terri's garden measures 24 feet by 36 feet. What is the area of her garden?

18.  January has 31 days. Since there are 24 hours in a day, how many hours are in the month of January?

# LESSON TEST

Regroup and multiply.

1.  $\begin{array}{r} 3\,1\,2 \\ \times\,5\,3 \\ \hline \end{array}$

2.  $\begin{array}{r} 2\,5\,8 \\ \times\,7\,8 \\ \hline \end{array}$

3.  $\begin{array}{r} 4\,7\,5 \\ \times\,7\,6 \\ \hline \end{array}$

4.  $\begin{array}{r} 3\,1\,6 \\ \times\quad 5 \\ \hline \end{array}$

5.  $\begin{array}{r} 3\,4 \\ \times\,9\,6 \\ \hline \end{array}$

6.  $\begin{array}{r} 8\,1 \\ \times\,1\,1 \\ \hline \end{array}$

Find the area and perimeter.

35 ft

12 ft    12 ft

35 ft

18 in

15 in    15 in

18 in

7.  Area = _____

8.  Perimeter = _____

9.  Area = _____

10.  Perimeter = _____

Write <, >, or = in the oval.

11.  $7 \times 2$ ◯ $7 + 2$        12.  $6 \times 6$ ◯ $4 \times 9$

13.  $9 \times 5$ ◯ $36 + 4$

Skip count and write the missing numbers.

14.  _____, 12, _____, _____, _____, _____, _____, _____, 54, _____

15.  Erin bought a piece of fabric measuring 36 inches wide by 108 inches long. What was the area of the fabric?

16.  Debbie earned $250 a week for a year (52 weeks). What were her total year's earnings?

17.  Shawn hauled 625 bags of corn. The bags each weighed 55 pounds. What was the total weight of the corn on Shawn's truck?

18.  Kent delivered 137 newspapers every day for 31 days. How many newspapers did he deliver altogether?

Find all the possible pairs of factors.

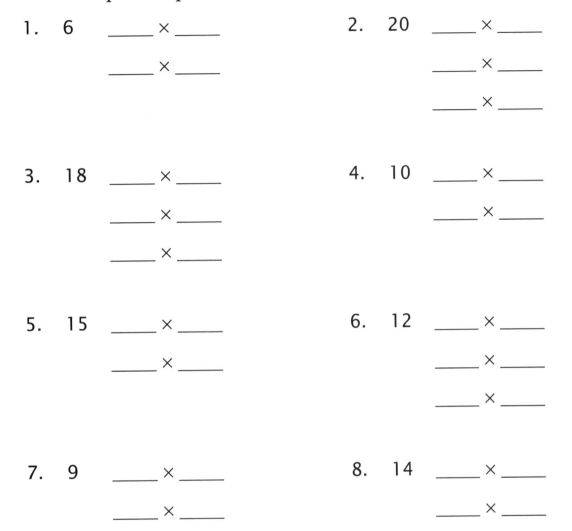

1.  6    ____ × ____

    ____ × ____

2.  20    ____ × ____

    ____ × ____

    ____ × ____

3.  18    ____ × ____

    ____ × ____

    ____ × ____

4.  10    ____ × ____

    ____ × ____

5.  15    ____ × ____

    ____ × ____

6.  12    ____ × ____

    ____ × ____

    ____ × ____

7.  9    ____ × ____

    ____ × ____

8.  14    ____ × ____

    ____ × ____

Multiply and fill in the blanks.

9.  3 quarters = _____ cents

10.  14 quarters = _____ cents

Regroup and multiply.

11.  1 8 5
     × 1 3

12.  6 9 3
     × 2 1

13.  5 6 4
     × 4 8

Fill in the blanks with the correct numerators and denominators to name the equivalent fractions.

14.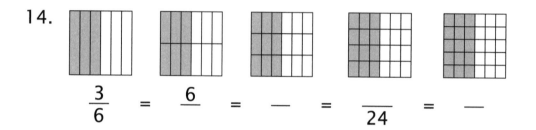

$\dfrac{3}{6}$ = $\dfrac{6}{\phantom{0}}$ = $\dfrac{\phantom{0}}{\phantom{0}}$ = $\dfrac{\phantom{0}}{24}$ = $\dfrac{\phantom{0}}{\phantom{0}}$

15. How many unit blocks are in a rectangle four units long and six units wide?

16. Briana has 23 quarters. How many cents does she have?

17. Randy picked out 16 unit blocks. List all the ways he can arrange them to form a rectangle.

_____ × _____ , _____ × _____ , _____ × _____

18. Jill wants to buy a birthday card that costs 69¢. Will three quarters be enough to pay for the card?

Write each number in words.

1. 6,701,413 _____

_____

2. 570,348 _____

_____

Write each number in standard notation.

3. 400,000 + 2,000 + 500 + 10 + 9 _____

4. 100,000,000 + 70,000,000 + 9,000,000 + 400,000 + 50,000
   + 7,000 + 300 + 80 + 5 _____

5. 200,000 + 4,000 + 200 + 10 + 3 _____

Write each number in place-value notation.

6. 975,236,759 _____

_____

7. 342,776 _____

_____

Multiply to find how many ounces.

8. 19 lb = _____ oz            9. 47 lb = _____ oz

10. 4 lb = _____ oz

Find all the possible pairs of factors.

11.  16  _____ × _____           12.  20  _____ × _____

         _____ × _____                    _____ × _____

         _____ × _____                    _____ × _____


13.  8   _____ × _____           14.  6   _____ × _____

         _____ × _____                    _____ × _____


Find the missing multiples of 2 and 9 in the equivalent fractions.

15.  $\dfrac{2}{9}$ = —— = —— = —— = —— = $\dfrac{}{54}$ = —— = $\dfrac{16}{}$ = —— = ——

16. Roxanne bought a 20-pound bag of flour. How many ounces of flour does she have?

17. Julia won a prize for her 36-pound pumpkin. Give her pumpkin's weight in ounces.

18. Scott caught a fish weighing 26 pounds. How many ounces does Scott's fish weigh?

Multiply.

1.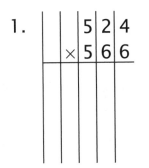

$$\begin{array}{r} 5\,2\,4 \\ \times\ 5\,6\,6 \\ \hline \end{array}$$

2.

$$\begin{array}{r} 8\,8\,6\,7 \\ \times\ \ \ \ 9\,3 \\ \hline \end{array}$$

3.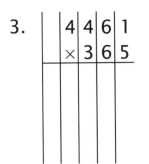

$$\begin{array}{r} 4\,4\,6\,1 \\ \times\ \ 3\,6\,5 \\ \hline \end{array}$$

4. Write in standard notation:

5,000,000 + 100,000 + 30,000 + 7,000 + 200 + 10 + 3

_____

5. Write in words:

44,900,000 _____

_____

6. Write in place-value notation:

517,058,800 _____

_____

Find all the possible pairs of factors.

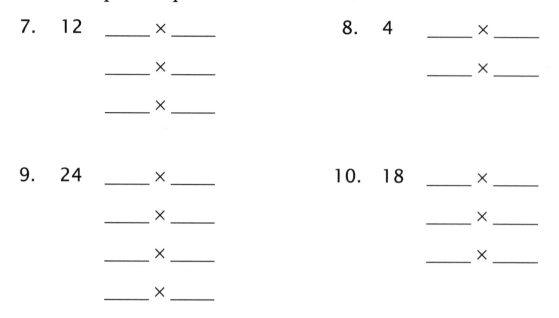

7.  12  ____ × ____

    ____ × ____

    ____ × ____

8.  4  ____ × ____

    ____ × ____

9.  24  ____ × ____

    ____ × ____

    ____ × ____

    ____ × ____

10.  18  ____ × ____

    ____ × ____

    ____ × ____

Solve for the unknown.

11.  6A = 48

12.  9Q = 81

13.  7T = 49

14.  8Y = 56

15.  Ann has a square area fenced off for her horse. The space measures 125 feet on each side. How many square feet of space does her horse have roam?

16.  Mike's widget factory made a widget every second. How many widgets does the factory make in an eight-hour work day?

# LESSON TEST

List all possible pairs of factors and tell whether the number is prime or composite.

1. 6   ____ × ____

    ____ × ____

    _____

2. 18   ____ × ____

    ____ × ____

    ____ × ____

    _____

3. 23   ____ × ____

    _____

4. 15   ____ × ____

    ____ × ____

    _____

5. 19   ____ × ____

    _____

6. 12   ____ × ____

    ____ × ____

    ____ × ____

    _____

Fill in the blanks.

7. 10 ft = ____ in

8. 9 years = ____ months

9. 6 dozen = ____ pencils

Multiply.

10.
$$\begin{array}{r} 4\,8\,5 \\ \times\ 7\,1\,2 \\ \hline \end{array}$$

11.
$$\begin{array}{r} 5\,4\,9\,1 \\ \times\ 3\,6 \\ \hline \end{array}$$

12.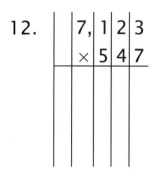

$$\begin{array}{r} 7{,}123 \\ \times\ 547 \\ \hline \end{array}$$

Add or subtract.

13.
$$\begin{array}{r} 892 \\ -173 \\ \hline \end{array}$$

14.
$$\begin{array}{r} 925 \\ +86 \\ \hline \end{array}$$

15.
$$\begin{array}{r} 204 \\ +139 \\ \hline \end{array}$$

16. List the ways you could make a rectangle that has an area of 24 blocks.

_____ × _____, _____ × _____, _____ × _____, _____ × _____

17. How many months had Keith lived when he reached his seventh birthday?

18. Ralph gathered six dozen eggs from his hen house. How many eggs did he have?

# LESSON TEST

Fill in the blanks.

1.  1 mile = _____ feet

2.  1 ton = _____ pounds

Multiply and fill in the blanks.

3.  3 miles = _____ feet

4.  22 tons = _____ pounds

5.  7 miles = _____ feet

6.  3 tons = _____ pounds

7.  12 quarters = ___ pennies

8.  9 pounds = _____ ounces

9.  11 feet = _____ inches

10.  7 doz. eggs = _____ eggs

Multiply.

11.

$$\begin{array}{r} 3\,9\,2 \\ \times\ 2\,3\,4 \\ \hline \end{array}$$

12.
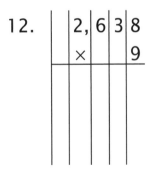
$$\begin{array}{r} 2{,}6\,3\,8 \\ \times\ \quad 9 \\ \hline \end{array}$$

13.
$$\begin{array}{r} 4{,}7\,0\,5 \\ \times\ \quad 2\,5 \\ \hline \end{array}$$

Write <, >, or = in the oval.

14. $8 \times 8$ ◯ $9 \times 7$          15. $3 \times 4$ ◯ $2 \times 6$

16. $5 \times 5$ ◯ $4 \times 7$

17. Adam's truck weighed 18 tons, and the load it was carrying weighed 22 tons. What was the weight of the truck and its load together, expressed in pounds?

18. Nate's airplane flight took him six miles above the earth's surface. How many feet above the ground was he?

19. Dave spread five tons of salt on the road with his snowplow. How many pounds of salt did he use?

20. Rhonda took a 14-mile ride on her bicycle. Express that distance in feet.

## IV

Regroup and multiply.

1.     2 1
    × 4 8

2.     3 6 4
    ×   5 3

3.     1 0 6
    × 7 8 9

4.  1, 3 5 7
    ×       6

5.  2, 8 4 3
    ×     7 5

6.  4, 5 6 1
    ×     3 2

Find all the possible pairs of factors and tell whether the number is prime or composite.

7.  9    _____ × _____

    _____ × _____

    _____

8.  12    _____ × _____

    _____ × _____

    _____ × _____

    _____

9.  7    _____ × _____

    _____

Fill in the blanks.

10. 8 quarters = _____ cents          11. 10 lb = _____ oz

12. 3 ft = _____ in          13. 6 miles = _____ feet

14. 2 tons = _____ lb          15. 25 ft = _____ in

16. What is 85 rounded to the nearest 10?

17. What is 105 rounded to the nearest 100?

18. What is 4,611 rounded to the nearest 1,000?

19. Write in standard notation: 4,000 + 500 + 60 + 8

_____

20. Write in place-value notation: 2,391,600 _____

_____

# FINAL TEST

Regroup and multiply.

1.      8 5
      × 2 6

2.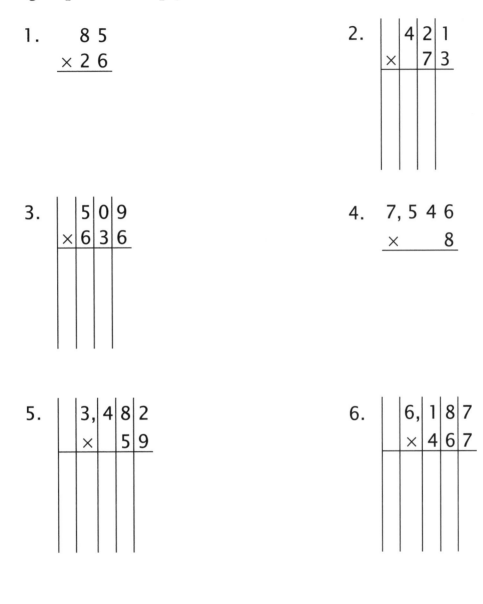
    | 4 | 2 | 1 |
    × | | 7 | 3 |

3.  | 5 | 0 | 9 |
    × | 6 | 3 | 6 |

4.  7, 5 4 6
        ×    8

5.  | 3, | 4 | 8 | 2 |
    × | | | 5 | 9 |

6.  | 6, | 1 | 8 | 7 |
    × | | 4 | 6 | 7 |

Find the area and perimeter.

7.  Area = _____

8.  Perimeter = _____

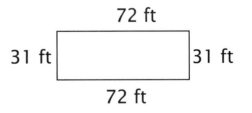
72 ft
31 ft      31 ft
72 ft

Solve for the unknown.

9. 8B = 64

10. 9Q = 63

11. 10X = 100

Find all the possible pairs of factors and tell whether the number is prime or composite.

12. 16 _____ × _____

_____ × _____

_____ × _____

_____

13. 7 _____ × _____

_____

14. 9 _____ × _____

_____ × _____

_____

Write <, >, or = in the oval.

15. 6 × 2 ◯ 3 × 4

16. 9 × 8 ◯ 5 × 12

17. 7 × 6 ◯ 9 × 5

Add.

18.
```
    9 2
    2 1
    4 8
  + 1 7
```

19.
```
    1 6 3
  +   5 4
```

20.
```
    8 1 5
  + 4 8 2
```

21.
```
    3 6 0
  -   3 7
```

22.
```
    5 2 9
  - 1 6 8
```

23.
```
    4 0 2
  - 2 9 3
```

Fill in the blanks.

24. 6 qt = _____ pt

25. 8 dimes = _____ cents

26. 9 yd = _____ ft

27. 5 Tbsp = _____ tsp

28. 10 nickels = _____ cents

29. 7 gal = _____ qt

30. $2 = _____ quarters

31. 4 gal = _____ pt

32. 3 lb = _____ oz

33. 6 quarters = _____ cents

34. 2 miles = _____ feet

35. 1 ton = _____ lb

36. A room measures 21 feet by 38 feet. Round the dimensions to the nearest ten and estimate the area of the room.

37. Chuck drove 452 miles a day for three days. Round to the nearest hundred and estimate how far he drove in all.

38. What is 3,495 rounded to the nearest thousand?

39. Write in standard notation: one million, two hundred seventy-one thousand, twenty-eight

_____

40. Write in place-value notation: 5,681,900

_____

_____